THE POETRY OF ANTIMONY

The Poetry of Antimony

Walter the Educator

SKB

Silent King Books a WhichHead Imprint

Copyright © 2023 by Walter the Educator

All rights reserved. No part of this book may be reproduced in any manner whatsoever without written permission except in the case of brief quotations embodied in critical articles and reviews.

First Printing, 2023

CONTENTS

Dedication v

Why I Created This Book? 1

One - Mystical Essence 2

Forever Concealed 4

Two - Serene And Pure 6

Three - Forever Unchained 8

Four - Igniting Reactions 10

Five - Story To Unfold 12

Six - Proclaim 14

Seven - Indelible Mark 16

Eight - Shining Bright 18

Nine - Metal Of Wonder 20

Ten - A Metal's Dance 22

Eleven - Exploring Your Mysteries 24

Twelve - Even In Darkness	. . . .	26
Thirteen - Gift And A World	. . . .	28
Fourteen - Beauty In The Shadows	. . . . . .	30
Fifteen - Wonder Hides	. . . .	32
Sixteen - Mysterious Spark	. . . .	34
Seventeen - Intricate Arc	. . . .	36
Eighteen - Embracing The Other	. . . . . .	38
Nineteen - Harmony In Complexity	. . . . .	40
Twenty - Enigmatic Realm Of Antimony	. . .	42
Twenty-One - Delicate Creed	. . . . . .	44
Twenty-Two - What Is And Could Be	. . .	46
Twenty-Three - True Beauty	. . . .	48
Twenty-Four - Eternal Gloom	. . . . . .	50
Twenty-Five - Cultivate	. . . .	52
Twenty-Six - Each Day	. . . .	54
Twenty-Seven - Power Within	. . . .	56
Twenty-Eight - Antimony Holds The Key	. . .	58
Twenty-Nine - Dance Of Shadows	. . . . .	60
Thirty - Eager To Share	. . . .	62
Thirty-One - Antimony's Allure	. . . .	64

Thirty-Two - Tread With Caution	. . . . . .	66
Thirty-Three - Behest	. . . .	68
Thirty-Four - Night And Day	. . . .	70
Thirty-Five - Testament	. . . .	72
About The Author	. . . .	74

WHY I CREATED THIS BOOK?

Creating a poetry book about the chemical element of Antimony was a unique and intriguing endeavor. Antimony, with its fascinating properties and historical significance, provides a rich source of inspiration. Its symbolism, atomic structure, and the role it has played in various cultures and industries can be explored through poetry, offering a fresh perspective on this element. By delving into the characteristics and stories associated with Antimony, I can create a body of work that educates, entertains, and captivates readers with its blend of science and art.

ONE

MYSTICAL ESSENCE

In the depths of nature's clandestine realm,
Lies a metal with an allure like a gem.
Antimony, the mysterious element it may be,
An enigma of the periodic table, you see.

A lustrous shadow, a silvery sheen,
With a touch of darkness, a haunting gleam.
Unyielding and steadfast, in its atomic embrace,
Antimony reveals its enigmatic face.

Born from the womb of the Earth's embrace,
In stibnite veins, it finds its resting place.
A symbol of transformation, of ancient alchemy,
Antimony's secrets whisper through history.

Its name derived from Greek, "anti-monos" they say,
A metal that opposes, in an arcane way.

A paradoxical character, both friend and foe,
With medicinal virtues, yet a poisonous blow.
    In alloys and compounds, it finds its use,
In pyrotechnics and cosmetics, it holds its dues.
A catalyst, a flame-retardant, a potent ally,
Antimony, the enigmatic metal, never wry.
    From the depths of the Earth to the laboratory's domain,
Antimony's allure continues to reign.
A symbol of the unknown and the unexplained,
In its mystical essence, it shall forever remain.

FOREVER CONCEALED

In the realm of elements, there lies a mystery untold,
A silvery sheen with a touch of darkness, behold!
Antimony, the enigma, born in Earth's embrace,
An ancient alchemist's dream, a puzzle to chase.

Friend and foe, this paradoxical entity,
A tonic for ailments, yet a poison of toxicity.
Medicinal virtues and poisonous allure,
Antimony, a paradox that will forever endure.

In alloys it finds solace, strength it imparts,
With lead and tin, a fusion of hearts.
Pyrotechnics' delight, it paints the sky with fire,
A blazing spectacle, a mesmerizing desire.

Cosmetics it adorns, with a touch of grace,
Enhancing beauty, leaving a trace.
A catalyst it becomes, igniting reactions,
A flame-retardant, safeguarding our creations.

Antimony, a symbol of the unknown and unexplained,
A riddle unsolved, a secret unrestrained.
Through the sands of time, its allure persists,
A silent witness to the mysteries that exist.

Oh Antimony, you captivate with your charm,
Unveiling secrets, causing hearts to disarm.

In your depths lies wisdom, waiting to be revealed,
A tale of fascination, forever concealed.

TWO

SERENE AND PURE

In the realm of elements, a mystic hue,
Lies Antimony, enigma anew.
A paradox it is, both dark and bright,
With traits that dance between day and night.
　Medicine's servant, a healer's balm,
It battles ailments with a gentle calm.
Yet in alloys' embrace, it finds its worth,
Strengthening metals, giving them birth.
　In pyrotechnic displays it ignites,
A shimmering dance of dazzling lights.
But in cosmetics' realm, it lends its grace,
Enhancing beauty, adorning the face.
　Antimony, a riddle, a puzzle untold,
Its secrets hidden, its mysteries unfold.
A catalyst of change, it sparks reactions,
Igniting transformations, forging new factions.

Its allure captivates, drawing us near,
A mesmerizing force, both far and near.
A symbol of the unknown, unexplained,
Wisdom waiting, yet to be unchained.

Antimony, oh element profound,
In your enigmatic depths, wisdom is found.
Through time's eternal dance, you endure,
A silent sentinel, serene and pure.

THREE

FOREVER UNCHAINED

In the depths of the earth, Antimony lies,
A mysterious element, with secrets it hides.
Both friend and foe, it dances in disguise,
A paradoxical essence, that mesmerizes.

In ancient times, alchemists revered,
Its transformative power, they sought to adhere.
Medicinal elixirs, they would create,
Antimony's healing touch, they did venerate.

In alloys and pyrotechnics, it found its place,
Enhancing strength, with a fiery embrace.
Its presence in cosmetics, a touch of allure,
Adorning faces, with a touch so pure.

Yet, Antimony's essence, cannot be contained,
For it ignites reactions, both wild and untamed.

A catalyst of change, it sparks the unknown,
Transforming elements, into something grown.
    Oh Antimony, enigmatic and rare,
A symbol of wisdom, waiting to be shared.
Through darkness and light, you dance with grace,
Revealing secrets, in your silent space.
    So let us marvel at your paradoxical ways,
As you weave through the fabric, in infinite arrays.
Antimony, a mystic force, forever unexplained,
With wisdom yet to be revealed, forever unchained.

FOUR

IGNITING REACTIONS

In the realms of science, a mystery unfolds,
Antimony, the enigma, its tale yet untold.
A paradoxical element, both bless and curse,
With virtues medicinal, and allure perverse.

A metalloid, it dances in the periodic dance,
Symbol of Sb, its cosmic romance.
From the depths of Earth, it emerges strong,
An ancient secret, shrouded for so long.

With healing hands, it aids the ailing soul,
In potions brewed, it plays a vital role.
Antimony, the alchemist's cherished prize,
In remedies forged, its wisdom lies.

But beware the allure of this silver beast,
For in its veins, a poison does it unleash.

A toxic whisper, a venomous sting,
A double-edged sword, this curious thing.
   In alloys forged, it lends its strength,
A catalyst of change, it goes to any length.
In pyrotechnics, it dazzles with its light,
A spark of wonder, igniting the night.
   And in the realm of beauty, it finds its place,
In cosmetics grand, it adds its grace.
A touch of glamour, a touch of charm,
In Antimony's embrace, beauty takes form.
   Oh, Antimony, element of wonder and might,
A hidden gem, shining in the night.
With transformative power, you lead the way,
Igniting reactions, sparking change every day.

FIVE

STORY TO UNFOLD

In shadows deep, where secrets dwell,
There lies a metal, Antimony's spell.
A paradox, enigma rare,
Its mysteries hidden, beyond compare.

A lustrous beauty, with a silver sheen,
Antimony, the alchemist's dream.
In alloys fused, its strength defined,
An ancient art, through time refined.

Pyrotechnics dance in fiery delight,
As Antimony's flames ignite the night.
Sparkling stars in the midnight sky,
A spectacle, captivating the eye.

Cosmetics adorned with its gentle touch,
A powder fine, that enhances much.

A blush of pink, a hint of grace,
Antimony's allure, shining in every face.

    But beyond the surface, a deeper role,
Antimony's secrets, a catalyst for change.
In medicine's embrace, it finds its place,
A healer's touch, with power to erase.

    Yet caution whispers, a warning's plea,
For Antimony's touch, can cause decree.
In shadows deep, where dangers lie,
It beckons us closer, with a mystifying sigh.

    Oh, Antimony, enigmatic and profound,
A silent force, forever unbound.
A metal of wonders, a mystery untold,
In your essence, a story to unfold.

SIX

PROCLAIM

In the depths of ancient times, a metal did arise,
With a paradoxical nature that bewitched all eyes.
Antimony, the enigma, its secrets yet untold,
A substance both alluring and poisonous, we're told.

    In the annals of history, its presence does unfold,
From the alchemists' lore to the mines of old.
Symbol of transformation, it holds a mystic sway,
As it weaves through the tapestry of night and day.

    In pyrotechnic displays, it dances in the flames,
Whispering secrets, playing dangerous games.
Its sparks ignite the heavens, a celestial ballet,
As it paints the sky with hues of vibrant display.

    Cosmetics it graces, with a touch of allure,
Enhancing beauty, a secret it ensures.

Adorning faces with a radiant glow,
Antimony's charm, in every stroke, does show.
    A healer in disguise, it mends the broken soul,
With remedies potent, making the spirit whole.
Yet beware its venom, for it can bring demise,
A double-edged sword, with secrets it implies.
    Antimony, the paradox, a metal so profound,
In its intricate essence, mysteries abound.
From poison to remedy, from allure to flame,
Its transformative power, forever shall proclaim.

SEVEN

INDELIBLE MARK

In pyrotechnics, Antimony shines,
A catalyst for sparks and fiery lines.
With its atomic number fifty-one,
It dances in the sky, its work undone.
    In cosmetics, Antimony finds its way,
Adorning faces in a subtle display.
A touch of allure, a hint of grace,
Its presence brings beauty to every face.
    But beware, for Antimony has a sting,
A toxic embrace that danger can bring.
Its poisonous touch, a double-edged sword,
A substance both transformative and untoward.
    In medicine, Antimony plays its part,
A healing agent with a magical art.

With remedies brewed, ailments it mends,
Yet its potency requires careful amends.
    Oh, Antimony, elemental star,
A paradoxical substance both near and far.
Igniting the sky with your dazzling light,
Enhancing beauty, both day and night.
    A symbol of change, a catalyst of fate,
Antimony, you captivate, allure, and sedate.
With danger and allure intertwined,
You leave an indelible mark on the human mind.

EIGHT

SHINING BRIGHT

In pyrotechnics, Antimony gleams,
A fiery dance, a dreamer's screams.
Bursts of color, sparks that ignite,
In the darkest of hours, a dazzling light.

    Cosmetics adorned with Antimony's touch,
A subtle allure, a whispering hush.
Eyes aglow, with a mystical gaze,
An enchanting spell, that never betrays.

    A catalyst for change, Antimony's might,
Unleashing reactions, both day and night.
Transforming the ordinary, with a magical sway,
Merging elements, in a wondrous display.

    Now a new tale, I shall unfold,
Of Antimony's nature, both fierce and bold.

Toxic it may be, in its lethal embrace,
Yet a healer it becomes, with grace.
    In medicine's realm, it finds its way,
A cure for ailments, a hope in dismay.
Balancing the scales, of life and death,
Antimony's touch, a fragile breath.
    Oh, Antimony, a paradox you are,
A seductive allure, a dangerous star.
Beauty and danger, entwined in your core,
An enigma, forevermore.
    In every facet, you captivate,
A substance, so mysterious, so innate.
Antimony, the alchemist's delight,
A symphony of elements, shining bright.

NINE

METAL OF WONDER

In the realm of elements, a paradox unfolds,
Antimony, a metal, with secrets untold.
Toxic and healing, a dual nature it bears,
An enigmatic essence, that captivates and ensnares.

A poison it may be, with venomous might,
Yet within its depths, a healing touch takes flight.
In ancient times, its properties were revered,
A remedy for ailments, when all else appeared.

In cosmetics it dances, a subtle allure,
Enhancing beauty, with a touch so pure.
Bathed in its essence, faces aglow,
A mesmerizing charm, it's beauty to bestow.

Pyrotechnics ignite, a symphony of fire,
Antimony's role, a dazzling desire.
In flames it dances, with brilliance and might,
A spectacle of light, captivating the night.

But beware, dear souls, of its potent embrace,
For Antimony's power, demands cautious grace.
A transformative force, both curse and boon,
A whisper of danger, in its alluring tune.

Oh, Antimony, a paradox so rare,
Both poison and healer, a tale to share.
In your essence lies a mystery, profound,
A metal of wonder, forever unbound.

TEN

A METAL'S DANCE

In shadows deep, a gleaming hue,
Antimony's beauty, both old and new.
A metal rare, mysterious and bold,
Its secrets whispered, yet untold.

A muse to alchemists, seekers of gold,
Antimony's allure, a story untold.
From earth's embrace, it's gently mined,
To forge a metal, as if refined.

In cosmetics, it finds its way,
A touch of glamour, a hint of sway.
Adorning eyes, with a smoky grace,
Antimony's allure, upon a face.

In medicine's grasp, it lends a hand,
A healer, a balm, in a toxic land.
A dual nature, both poison and cure,
Antimony's touch, so pure, so pure.

On pyrotechnic nights, it dances bright,
Illuminating darkness, with fiery might.
A spark of passion, a dazzling show,
Antimony's light, in pyres aglow.

Yet danger lurks, within its core,
A double-edged sword, forevermore.
A cautionary tale, of power unleashed,
Antimony's enchantment, never to be ceased.

Oh Antimony, enigmatic and rare,
You captivate hearts, beyond compare.
In your essence lies, a timeless art,
A metal's dance, within the human heart.

ELEVEN

EXPLORING YOUR MYSTERIES

In the realm where shadows dance and flicker,
There lies a substance, mysterious and thicker.
Antimony, the enigmatic alchemist's dream,
With allure and danger, its essence doth teem.

A paradoxical presence, both dark and light,
A metal, a poison, a shimmering sight.
Its atomic number, fifty-one, holds the key,
Unlocking the secrets of its chemistry.

An alluring luster, like moonlight on the sea,
Antimony bewitches, capturing hearts with glee.
Its crystalline structure, sharp and precise,
Reflects its dual nature, both virtue and vice.

Beneath the surface, a potent elixir brews,
A touch of madness, a cure, or the muse.

For in its depths, transformations unfold,
A catalyst for change, a story yet untold.
    But heed this warning, ye seekers of might,
For Antimony's touch can blind in the night.
Its toxic embrace, a perilous delight,
A dance with danger, a treacherous flight.
    Oh, Antimony, the paradox profound,
A substance of wonders, both lost and found.
With caution and reverence, we tread this path,
Exploring your mysteries, embracing the aftermath.

TWELVE

EVEN IN DARKNESS

In the depths of the earth, where shadows dwell,
Lies a secret element, Antimony's spell.
A paradox it is, both beauty and bane,
A substance that can heal, yet cause much pain.

Its shimmering allure, a silver sheen,
Captivating hearts with an enigmatic gleam.
But hidden beneath its captivating guise,
Lies a toxicity that can torment and agonize.

An alchemist's dream, a catalyst of change,
Antimony weaves its magic, rearranging the strange.
In fiery pyrotechnics, it dances with delight,
Igniting the sky with a mesmerizing light.

Yet beware, oh mortal, of its poisonous sting,
For Antimony's touch can be a deadly thing.

A delicate balance, a tightrope it walks,
Between life and death, where the fragile heart talks.
    Oh Antimony, you hold a power untamed,
A paradoxical element, forever unnamed.
You enchant and harm, you heal and destroy,
Transforming the world with your paradox of joy.
    So let us marvel at your enigmatic grace,
As we navigate the mysteries of this cosmic space.
For in your essence, we find the truth divine,
That even in darkness, a silver lining can shine.

THIRTEEN

GIFT AND A WORLD

In the depths of earth, where shadows reside,
There lies a metal, with secrets to confide.
Antimony, they call it, a mysterious name,
With a dual nature, both a bane and a flame.

Toxic whispers dance in its silver sheen,
A poison that taints, a venom unseen.
Yet hidden within, a paradox unfolds,
For healing hands, Antimony beholds.

An alchemist's dream, a remedy's grace,
It mends the broken, with its potent embrace.
A salve for the soul, a cure for the mind,
Antimony's touch, a solace to find.

But beware, dear hearts, of its treacherous guise,
For Antimony's allure can mesmerize.

In pyrotechnic displays, it takes center stage,
Bathing the night sky in a fiery rage.

Its sparks ignite, like stars ablaze,
A spectacle captivating, in crimson haze.
But in its brilliance, a warning lies,
For Antimony's fire can burn too bright.

Oh, Antimony, enchanting yet fierce,
With powers to heal and powers to pierce.
A metal of contradictions, a paradox untamed,
Both a savior and a bane, forever unnamed.

So let us tread lightly, in Antimony's realm,
Respecting its power, its secrets to overwhelm.
For within its essence, a tale is unfurled,
Of a chemical element, both a gift and a world.

FOURTEEN

BEAUTY IN THE SHADOWS

In the realm of elements, a force emerges,
A healer and a danger, entwined in one,
Antimony, with its touch that can be pure,
Yet toxic, a paradox beneath the sun.

A silvery hue, like moonlight in the night,
It whispers secrets, veiled in mystic grace,
With alchemical power, it takes its flight,
Transforming souls, leaving no trace.

Antimony, a herald of pyrotechnics,
A dazzling light, captivating and bright,
In fiery displays, its essence flicks,
A dance of flames, igniting the night.

But tread with caution, for danger lies,
In Antimony's allure, a treacherous game,

For in its depths, a darkness hides,
A touch that can both heal and maim.
    So let us marvel at its enigmatic grace,
Embrace its power, but with wary eyes,
For Antimony, in its evocative embrace,
Can enchant and harm, heal and destroy.
    In this element's essence, a silver lining gleams,
A reminder that even darkness holds a spark,
For in the dance of Antimony's dreams,
We find beauty in the shadows, and hope in the dark.

FIFTEEN

WONDER HIDES

In shadows deep, where darkness dwells,
There lies a secret, tales it tells.
Antimony, with its mystic guise,
A paradox, enchanting, wise.

A metal rare, its beauty gleams,
A silver hue, like moonlit dreams.
Yet in its heart, a venom brews,
A potent elixir, one must choose.

A healer's touch, a soothing balm,
Antimony's power can disarm.
It mends the body, mends the soul,
Restoring balance, making whole.

But heed the warning, tread with care,
For Antimony's allure, a snare.
Its touch, a poison, lethal, cold,
A silent killer, ruthless, bold.

A dance of light, a dance of night,
Antimony's symphony, pure delight.
Its flames ignite, a fiery trance,
A flickering glow, a devil's dance.

A pendulum swings, between the two,
Antimony's duality, a clue.
To know its secrets, to understand,
Requires wisdom, a steady hand.

So let us marvel, let us explore,
Antimony's depths, forevermore.
In light and darkness, it resides,
A paradox, where wonder hides.

SIXTEEN

MYSTERIOUS SPARK

In Antimony's realm, a paradox unfolds,
A dual nature it holds, both light and dark, I'm told.
A metalloid, mysterious and rare,
With secrets whispered in its atomic lair.
 A healer it claims to be, with potent might,
Binding wounds, curing ailments, bringing light.
But beware of its touch, a venomous sting,
For in its depths, a poison it does bring.
 Like a chameleon, it changes its form,
Shifting between solid and liquid, a transformative swarm.
A silvery sheen, gleaming bright,
Yet concealing a shadow, a hidden blight.
 Oh, Antimony, you mesmerizing flame,
With an allure that's impossible to tame.

Burning fiercely, casting a spell,
Yet leaving scars, a tale to tell.

Balance is key, in your presence we must tread,
With caution and respect, or we'll be misled.
For you hold the power to heal and destroy,
To captivate and harm, oh Antimony, oh alloy.

A dance with danger, a delicate art,
To harness your essence, and tame your heart.
In your enigmatic presence, we find,
A lesson profound, within the confines.

Antimony, oh chemical enigma,
Teach us the wisdom, the delicate stigma,
Of embracing contradictions, both light and dark,
And finding harmony in your mysterious spark.

SEVENTEEN

INTRICATE ARC

In the depths of the earth's embrace,
Lies a flame of mesmerizing grace.
Antimony, the element of allure,
With powers to heal and powers to obscure.
   A flickering fire, like a siren's call,
It dances and tempts, standing tall.
Beware its beauty, its radiant glow,
For within lies a paradoxical flow.
   With alchemical secrets, it weaves its spell,
Unveiling mysteries as it compels.
Healer of pain, mender of wounds,
Yet harbinger of chaos, destroyer of tunes.
   In the hands of the wise, a potent elixir,
A catalyst for change, a bold fixer.
Yet in the hands of the reckless, a venomous brew,
A poisonous potion, a dangerous stew.

Antimony, a lesson in duality,
A reminder of life's inherent complexity.
For it teaches us to embrace contradiction,
To find harmony amidst the friction.

Balance and caution, the keys to its essence,
A dance with fire, a delicate presence.
Let its lessons guide us, both near and far,
To navigate the paradoxes that are.

Oh, Antimony, enigmatic and bold,
A tale of wonder that will forever be told.
In your flame, we find both light and dark,
A lesson in life's most intricate arc.

EIGHTEEN

EMBRACING THE OTHER

In the realm of elements, there lies Antimony,
A flame so mesmerizing, its allure impossible to flee.
A dance of shadows in the alchemist's art,
A substance both healing and tearing us apart.
   Beneath its surface, a duality resides,
A paradoxical nature that no one can hide.
With every touch, a spell is cast,
A delicate balance, a power unsurpassed.
   Antimony, the enigma, whispers ancient tales,
Of secrets unlocked and forbidden trails.
It holds the key to both life and death,
A paradoxical elixir, with every breath.
   With wisdom and care, its essence we must wield,
For Antimony's power can both harm and heal.

In its embrace, a lesson we find,
To navigate contradictions, and balance we bind.
    So let us tread lightly on this treacherous path,
Embracing the contradictions, avoiding the aftermath.
For in Antimony's flame, a truth we uncover,
That life's essence lies in embracing the other.

NINETEEN

HARMONY IN COMPLEXITY

In Antimony's realm, a paradox unfolds,
A dance of light and darkness, a story untold.
For in its essence, both venom and cure reside,
A potent elixir, where caution must abide.

Its touch can heal, with powers profound,
But wielded carelessly, chaos will abound.
A double-edged sword, both ally and foe,
A lesson in balance, that Antimony bestows.

With a regal allure, it captures the eye,
An enigma of nature, none can deny.
Yet beneath its luster, a cautionary tale,
A reminder of harmony, that we must unveil.

For Antimony whispers secrets of old,
Of duality and wisdom, its essence bold.

It teaches us to embrace the contradictions we find,
To seek harmony amidst the chaos, entwined.

In its presence, caution must be our guide,
A reminder to navigate with wisdom beside.
For Antimony's power is not to be tamed,
But respected and harnessed, in balance acclaimed.

So let us learn from this enigmatic element,
To embrace the paradox, with minds fervent.
In Antimony's essence, a lesson profound,
To find harmony in complexity, we are bound.

TWENTY

ENIGMATIC REALM OF ANTIMONY

In the depths of Earth's embrace, Antimony lies,
A paradox of power, both gentle and wise.
A metal of mysteries, it dances with flame,
Whispering secrets, unknown, untamed.

On the alchemist's table, it weaves its spell,
Transforming the ordinary, in its presence, all dwell.
With magical touch, it heals and it harms,
A double-edged sword, with elusive charms.

A guardian of balance, it teaches us well,
In its essence, the story of heaven and hell.
For in its darkness, a flicker of light,
Reflecting the shadows, revealing insight.

Its poison, a venom, can bring pain and strife,
Yet in measured doses, it saves many a life.

A paradoxical guardian, both bane and friend,
Its nature, a riddle, with no clear end.
 Oh, Antimony, enigma, in your depths we see,
The dance of duality, the essence of thee.
Embrace contradictions, find harmony's key,
In the enigmatic realm of Antimony.

TWENTY-ONE

DELICATE CREED

In the realm of shadows, where darkness resides,
There lies a metal, mysterious and wise.
Antimony, they call it, a paradox untamed,
A catalyst for change, a force unnamed.

Its touch is poison, its essence so pure,
Both healer and destroyer, a dance obscure.
From the depths it emerges, a teacher profound,
Showing us the balance, where secrets are found.

In the alchemist's hands, it transforms with grace,
From solid to liquid, a mystical embrace.
Its flames burn bright, a flickering light,
Guiding seekers through the depths of the night.

Beware, oh wanderer, its power untamed,
For Antimony's embrace can leave you maimed.

With caution and wisdom, we must proceed,
For the enigmatic metal holds the key.
    Embrace the contradictions, the duality within,
For life's lessons lie in the dance of sin.
In Antimony's presence, we learn to discern,
The delicate balance between heal and burn.
    So let us respect this element divine,
For in its paradox, true wisdom we find.
Antimony, the enigma, forever we'll heed,
A reminder that life is a delicate creed.

TWENTY-TWO

WHAT IS AND COULD BE

In Antimony's realm, a paradox dwells,
A chemical dance where healing compels,
Yet lurking beneath its gentle guise,
A power to harm, to mesmerize.

A silvery metal, a tender embrace,
With properties rare, a mystic space,
An alchemist's dream, a sorcerer's brew,
Antimony's essence, both old and new.

With caution we tread, in this enigmatic land,
For Antimony's touch, a delicate hand,
It can mend broken hearts, heal wounds deep,
Or sow seeds of poison, secrets to keep.

A balance is needed, a delicate sway,
To harness its power, in an elegant way,

For in its duality, wisdom resides,
Teaching us to embrace contradictions, abide.
    Antimony, the teacher, whispers profound,
In its essence, lessons to be found,
Seek harmony amidst chaos, it imparts,
To navigate complexity, with caution and smarts.
    So let us respect this enigmatic friend,
Handle with care, its messages transcend,
For in Antimony's essence, a truth we see,
The delicate balance between what is and could be.

TWENTY-THREE

TRUE BEAUTY

In the depths of the earth, where darkness resides,
There lies a secret, where Antimony hides.
A paradox it is, this element of strife,
Both a bane and a blessing, in the dance of life.

With a silvery sheen, it lures the unsuspecting,
A beacon of allure, its power infecting.
But beware, oh mortal, of its devious ways,
For Antimony's touch can lead astray.

A dual nature it possesses, a double-edged sword,
Healer and destroyer, a balance restored.
In potions and elixirs, it mends the ailing,
Yet in the wrong hands, chaos it's unveiling.

A messenger of change, it whispers in the wind,
Unveiling truths hidden, where shadows have been.

An alchemist's dream, a philosopher's stone,
Antimony's wisdom, to the seeker it's shown.
    But tread with caution, for its power is vast,
Harnessing its essence, a delicate task.
For in the dance of life, where light meets dark,
Antimony's lessons leave a lasting mark.
    So let us honor this enigmatic element,
Embrace its contradictions, with reverence and sentiment.
For in Antimony's essence, we find the key,
To balance and harmony, where true beauty will be.

TWENTY-FOUR

ETERNAL GLOOM

In the realm of enigma, an element resides,
Antimony, a paradox that nothing hides.
A metalloid, both friend and foe,
Its secrets to the world it doesn't easily show.

With a lustrous sheen, it catches the eye,
But its touch, beware, for it can deceive and lie.
Its nature, mysterious, in shades of gray,
Teaching us to embrace contradictions and find our way.

Antimony, the alchemist's delight,
A symbol of transformation, both day and night.
Its presence, a catalyst for change,
A whisper of secrets, hidden and strange.

From potions brewed in ancient lore,
To weapons forged in times of war,

Antimony's power, a double-edged blade,
Caution and respect, for its essence must be paid.
  Yet, in healing hands, it holds the key,
To remedies that set our spirits free.
An antidote to ailments that hold us tight,
Antimony's touch, a balm in the night.
  Oh, Antimony, a delicate dance,
A balance to strike, a chance to enhance,
The duality within, the light and the shade,
Approach with reverence, let wisdom be your aid.
  For in this element, a message lies,
Of change and growth, as time flies.
Embrace its contradictions, let harmony bloom,
Antimony, the messenger of transformation, in its eternal gloom.

TWENTY-FIVE

CULTIVATE

In Antimony's realm, a world unseen,
A paradox of power, both cruel and keen.
Its touch heals wounds, mends broken hearts,
Yet wields a darkness that fiercely imparts.

A metal rare, mysterious and pure,
Its essence, a teacher, both sage and allure.
With caution, we harness its potent might,
For balance is key in Antimony's light.

A messenger of change, it whispers in our ear,
A catalyst for growth, a harbinger of fear.
In its presence, we learn to tread with care,
For Antimony's power demands our prayer.

A delicate dance of shadow and gleam,
Its alchemy weaves a mystical dream.

With reverence and respect, we must approach,
Antimony's essence, a delicate coach.
　In war, it forges swords, a weapon so bold,
In healing, it mends, its touch pure as gold.
A force of transformation, both fierce and wise,
Antimony's secrets, they mesmerize.
　So let us embrace this enigmatic guest,
With wisdom and caution, we'll be blessed.
For Antimony holds the key to our fate,
A reminder that balance, we must cultivate.

TWENTY-SIX

EACH DAY

In Antimony's realm, duality resides,
A dance of light and shadows it confides.
A paradox of healing and poison,
This element's essence we must question.
    With strength and wisdom, it does unfold,
A tale of lessons, both new and old.
In balance it thrives, a delicate scale,
Teaching us caution, its secrets unveil.
    Its touch can mend, a soothing balm,
Yet in excess, it brings only harm.
A paradox we must navigate,
To harness Antimony's transformative state.
    Its alchemical power, a mystical force,
Guiding us on a transformative course.

Through trials and tribulations it leads,
To find harmony amidst chaos' seeds.

    Oh, Antimony, enigmatic and rare,
With lessons to teach, and wisdom to share.
Embrace its duality, the light and the shade,
For in its essence, balance is made.

    So, let us tread with reverence and care,
In the realm of Antimony, we must fare.
Seeking its lessons, we'll discover the way,
To find harmony and balance each day.

TWENTY-SEVEN

POWER WITHIN

In shadows deep, where secrets lie,
An enigma born beneath the sky.
Antimony, oh mysterious one,
A tale of power, yet caution won.

A metal cold, a lustrous sheen,
With paradoxes that lie between.
Healer and harm, both hand in hand,
A dance of grace on shifting sand.

With healing touch, you mend the soul,
Dispelling darkness, making whole.
But tread with care, for in your grasp,
Lies the potential for pain to clasp.

A catalyst, you bring forth change,
With alchemy, you rearrange.

Transforming lead to golden light,
Yet beware, for darkness takes its flight.

Oh Antimony, a paradox so grand,
In your presence, we must understand,
That balance and caution must prevail,
To harness your power, yet not to fail.

Embrace your contradictions, hold them near,
For in their union, wisdom appears.
With reverence and respect, we learn,
To seek harmony, to grow, to discern.

Antimony, oh element divine,
A reminder to tread each path with thine,
With balance and caution, we shall embrace,
The power within, the mysteries we trace.

TWENTY-EIGHT

ANTIMONY HOLDS THE KEY

In the realm of elements, Antimony I find,
A paradoxical substance, both cruel and kind.
Its nature enigmatic, its secrets concealed,
A dance of contradictions, its essence revealed.

A metal, yet brittle, a solid, yet strange,
With powers of transformation, it can rearrange.
A catalyst of change, a harbinger of fate,
Antimony, the alchemist's ultimate mate.

Beware, oh mortal, of its mercurial might,
For in its dual nature, lies both dark and light.
A symbol of warfare, its toxicity known,
Yet a healer it becomes, when carefully sown.

In battles it's wielded, in bullets and blades,
A weapon of destruction, a choice man-made.

But in the hands of healers, it mends and repairs,
A balm for the wounded, a remedy rare.

Antimony, the enigma, with lessons profound,
Teaching us balance, where harmony is found.
Embrace its contradictions, its power to transform,
A reminder to seek the light, amid the storm.

Within its depths lie secrets, both ancient and wise,
Lessons of growth, of change, in paradox lies.
So approach with reverence, handle with care,
For Antimony holds the key, to our destiny fair.

TWENTY-NINE

DANCE OF SHADOWS

In the depths of the earth, where secrets lie,
A metal of paradox, Antimony by name,
Both healer and poison, its powers untamed,
A dance of duality that none can deny.

With darkness it beckons, a shadowy lure,
A deadly elixir, a venomous brew,
Yet in measured doses, it can heal and renew,
A delicate balance, where caution is sure.

Its touch is transformative, a catalyst rare,
Alchemist's treasure, it changes the game,
From lead to gold, the transmutation of fame,
A mystery wrapped in a shroud of despair.

But heed the warning, oh seeker of might,
For Antimony's power can lead to demise,

In harmony and balance, its secrets arise,
A lesson profound, in the dance of the night.
　　Tread lightly, dear souls, in Antimony's embrace,
Respect its enigma, its wonders unfold,
For in its contradictions, wisdom takes hold,
An alchemical journey, in time and in space.
　　Embrace the paradox, the yin and the yang,
For in Antimony's essence, a truth is unveiled,
The path to harmony, where balance prevails,
A dance of shadows, where light finds its tang.

THIRTY

EAGER TO SHARE

Antimony, the paradox in the palm of nature,
A metal of contradictions, both healer and destroyer.
Its touch brings both harm and remedy,
A dance of chaos and harmony.

In warfare, its venomous arrows fly,
Piercing armor, causing foes to cry.
Yet in the alchemist's hands, it transforms,
Creating elixirs, where healing forms.

A symbol of transformation and growth,
Antimony teaches the value of both.
Like a phoenix rising from ashes burned,
It shows us how wisdom is earned.

But tread with caution, for its power is great,
Too much can seal a tragic fate.

Balance, oh seeker, is the key,
To harness Antimony's potency.
    Embrace the contradictions it presents,
For in them lies hidden wisdom's essence.
Learn from its duality, let it be your guide,
Unveiling truths that within you reside.
    Antimony, enigmatic and rare,
A reminder that life is not always fair.
Embrace its mysteries, with reverence and care,
And find the lessons it's eager to share.

THIRTY-ONE

ANTIMONY'S ALLURE

In the depths of the earth, a secret lies,
A paradoxical element, both gentle and wise.
Antimony, they call it, a metal arcane,
With transformative power, a catalyst's reign.
    A touch of its essence, a dance with the flame,
Unleashes its power, a force to reclaim.
Alchemy's gift, a potion of change,
But tread with caution, for danger's in range.
    A balance delicate, like life's fragile thread,
Antimony's lesson, in caution we're led.
From poison to cure, it traverses the line,
A healer's embrace, a destroyer's design.
    In shadows it lurks, a sorcerer's brew,
A paradox unraveling, both false and true.

Its touch can bring fortune, or lead to demise,
An enigma of nature, where wisdom lies.

Oh Antimony, guide us with your light,
Illuminate the path, through chaos and night.
With reverence we seek, your secrets concealed,
To find the harmony, in balance revealed.

So let us embrace, this element rare,
With awe and respect, its mysteries we share.
For within its paradox, wisdom does dwell,
Antimony's allure, a tale we must tell.

THIRTY-TWO

TREAD WITH CAUTION

In the depths of the earth, a hidden treasure lies,
A paradoxical metal, with enigmatic guise.
Antimony, they call it, a mystery untold,
With transformative powers, both gentle and bold.

A lustrous silver sheen, with a touch of grey,
It dances with fire, in a mystical display.
Alchemy's darling, a catalyst profound,
Unleashing reactions, with a subtle sound.

In the crucible's embrace, it weaves its spell,
Melting, transforming, a secret to tell.
A symbol of change, of transmutation's might,
Antimony emerges, from darkness to light.

But heed its warning, this metal of old,
For balance is key, its nature to behold.

A double-edged sword, it can heal or harm,
A cautionary tale, of power's charm.
From the depths of the earth, to the heights of the sky,
Antimony's wisdom, it beckons you nigh.
Embrace its contradictions, its secrets untold,
And find harmony's dance, within its stronghold.
So let us marvel at Antimony's grace,
A puzzle unsolved, in time and in space.
In its paradox lies a lesson profound,
To tread with caution, on hallowed ground.

THIRTY-THREE

BEHEST

In the realm where shadows dance,
Lies a metal with a mystic stance.
Antimony, they call thee,
A paradoxical entity.
    From the earth's hidden womb it springs,
With secrets untold, ancient things.
A bearer of both harm and cure,
With power that's both dark and pure.
    Oh Antimony, enigma divine,
Alchemy's gift, a potion of time.
A catalyst for transformation's might,
Unveiling wisdom, blinding light.
    With venomous touch, it can corrupt,
Imparting poison, chaos abrupt.

Yet in measured hands, it brings relief,
A healer's touch, a balm, a belief.

 A paradox, this metal rare,
In its duality, a lesson to bear.
For in life's journey, we often find,
That light and dark are intertwined.

 Balance, dear seeker, is the key,
In harnessing Antimony's decree.
With reverence and caution, tread the line,
Embrace both sides, the divine and malign.

 For in this dance of light and shade,
Through Antimony's power, wisdom is made.
Let transformation guide your quest,
And find harmony, in life's behest.

THIRTY-FOUR

NIGHT AND DAY

In shadows deep, where secrets hide,
A paradox dwells, Antimony's pride.
A metal rare, both gift and curse,
Its touch can heal, its touch can pierce.

With alchemist's gaze, we peer within,
To unravel the mysteries it conceals.
A teacher stern, it whispers truths,
Of transformation, both dark and true.

From Egypt's tombs to China's lands,
Its tale unfolds through human hands.
A remedy for ills, a poison's sting,
Antimony's touch, a delicate thing.

In potions brewed, it mends the weak,
But veer too close, and havoc it wreaks.

A catalyst of change, a catalyst of doom,
Antimony dances in an ethereal gloom.

Tread with caution, ye who dare,
To harness Antimony's potent flare.
With reverence, respect, and care,
Its secrets, fragile, we must bear.

For in this metal's enigmatic core,
Lies wisdom ancient, wisdom pure.
Balanced scales, a delicate dance,
Antimony's essence, a fleeting chance.

So let us learn from its timeless lore,
To balance power, to explore.
In Antimony's realm, we find our way,
A paradoxical guide, both night and day.

THIRTY-FIVE

TESTAMENT

In shadows deep, where secrets sleep,
A paradox of nature's keep,
There lies a metal, dark and rare,
With tales of healing and despair.

Antimony, enigmatic and bold,
A substance ancient, stories untold,
Its touch can mend, its touch can kill,
A dance of chaos, a harmony still.

With alchemy's touch, it finds its form,
A mystical cure, a venomous storm,
A touch of poison, a healer's balm,
A delicate balance, a sacred psalm.

In potions brewed, it holds the key,
Unveiling truths, both dark and free,

A catalyst of change, a flickering light,
Guiding the lost through the darkest of night.
    Yet heed with caution, this double-edged blade,
For Antimony's power can swiftly fade,
In balance we find its essence true,
A force to nurture, a force to subdue.
    Oh, Antimony, enigma divine,
Within your depths, wisdom we find,
Embrace your contradictions, your dual face,
A testament to life's intricate grace.

ABOUT THE AUTHOR

Walter the Educator is one of the pseudonyms for Walter Anderson. Formally educated in Chemistry, Business, and Education, he is an educator, an author, a diverse entrepreneur, and he is the son of a disabled war veteran. "Walter the Educator" shares his time between educating and creating. He holds interests and owns several creative projects that entertain, enlighten, enhance, and educate, hoping to inspire and motivate you.

Follow, find new works, and stay up to date
with Walter the Educator™
at WaltertheEducator.com

www.ingramcontent.com/pod-product-compliance
Lightning Source LLC
LaVergne TN
LVHW051958060526
838201LV00059B/3710